779.9

NENDEN SCH
WITHDRAWN

D1379896

World

TROTH WELLS AND CASPAR HENDERSON
FOREWORD BY DR CAROLINE LUCAS

OUR FRAGILE
World

THE BEAUTY OF A PLANET UNDER PRESSURE

 Thames & Hudson

Any copy of this book issued by the publisher as a paperback is sold subject to the condition that it shall not by way of trade or otherwise be lent, resold, hired out or otherwise circulated without the publisher's prior consent in any form of binding or cover other than that in which it is published and without a similar condition including these words being imposed on a subsequent purchaser.

First published in the United Kingdom in 2005 by
Thames & Hudson Ltd,
181A High Holborn
London WC1V 7QX

www.thamesandhudson.com

Text © 2005 Troth Wells and Caspar Henderson.
Images © 2005 the individual photographers and agencies.

All Rights Reserved. No part of this publication may be reproduced or transmitted in any form or by any means, electronic or mechanical, including photocopy, recording or any other information storage and retrieval system, without prior permission in writing from the publisher.

This book was conceived, designed and produced by
New Internationalist™ Publications Ltd
55 Rectory Road
Oxford OX4 1BW, UK
www.newint.org

Designed by: Ian Nixon

British Library Cataloguing-in-Publication Data
A catalogue record for this book is available from the British Library

ISBN-13: 978-0-500-51275-3
ISBN-10: 0-500-51275-2

Printed and bound in China

we have reached the critical moment in history when we must urgently decide what relationship we will have with the planet that sustains all known life. Will we continue with our short-sighted plunder, or will we take the positive steps necessary to secure our common future? Through extraordinary photographs this important book shows why the stakes are so high and time is so short'

Tony Juniper
Executive Director of Friends of the Earth

The exodus trail. Human rights organizations estimate that 2.5 million people have been killed in the long-running war in the Democratic Republic of Congo, not only from violence but also from disease and malnutrition. Many have been made homeless, their lands and animals destroyed.

Adelie penguin, Antarctica.

F R

AGILE

WORL

Mass transit train, Taiwan.

CONTENTS

Foreword by
Dr Caroline Lucas, MEP   14

Introduction – The world
we find; the world we make   16

PART ONE – THE INHERITANCE   38

Fire & transformation   40

The seas   52

Forest & swamp   62

Mountains & polar regions   72

Managed lands   82

PART TWO – THE LEGACY   94

People   96

City life   106

Climate change   118

Agriculture & farming   138

Rivers & estuaries   152

Back to the future   164

Campaigning organizations,
resources, photographs & credits   187

As a Green Party Member of the European Parliament, I often find myself arguing over the technical details of environmental legislation – the details of the EU's emissions trading scheme, say, or the maximum permitted levels of pesticide use.

Rarely do I get the chance to step back and see the bigger picture: the picture presented so stunningly in *Our Fragile World: the beauty of a planet under pressure*.

The world that it captures is a world of Aristotle's four elements: earth, water, fire and air. They are presented both at their delicate best, playing their essential part in the earth's diverse ecosystems, and at their destructive worst in what must surely be read as a call to arms for the global environmental movement.

By focusing our attention so precisely on our impact on the earth and bearing witness to the frequent discord between people and the world that supports us, it makes us realize the true extent of the damage we have wreaked. It highlights the urgency of the action we must take to prevent the loss, not just of human civilization as we know it, but of the species and ecosystems documented here.

Senior government advisors on both sides of the Atlantic have described climate change as the single biggest threat we face in the 21st century, with greater destructive capacity than terrorist attack or war. It is time for governments to consider our impact on the world as a security issue and invest at least the same amount of political capital and financial resources in environmental solutions as they currently do on military capacity.

The driving force behind real change is the commitment of a vocal and demanding mass movement which, over the last 30 years, has increasingly called for radical change to the way we interact with the earth. We might not have the luxury of another 30 years.

Unless we, the human race, act now, we risk going down in history as the species that spent all our time monitoring our own extinction – and the destruction of the world's fragile ecosystems – rather than preventing it. *Our Fragile World* catalogues some of what would be lost.

But as well as being a call to arms, this book is also a message of hope: it is not yet too late to avert environmental catastrophe, not yet too late to cease – and begin reversing – humankind's destructive impact, not yet too late to prevent the extinction of the Bengal Tiger, the Hawksbill Turtle or the Fijian Giant Sea Fan.

Extraordinary pictures from some of the world's best photographers make this an invaluable book that will become respected worldwide for its accessible and comprehensive treatments of some of the biggest issues in the struggle for sustainability and global justice.

Our Fragile World, conceived by New Internationalist, is an inspirational source for all of us involved in the struggle for a sustainable, just world. It adds both urgency and immediacy to our work, by showing the beauty and fragility of the world around us – and exactly what we stand to lose if we fail.

Dr Caroline Lucas, MEP

THE WORLD WE FIND,

THE WORLD WE MAKE

Pages 14-15
Tassili desert, Algeria.
Pages 16-17
Banaue rice terraces, Philippines. The terraces have been designated a UNESCO Endangered World Heritage Site. They suffer due to erosion and the effects of the tourism industry.

▲ **Goldfish reef:** Coral is made of thousands of tiny marine animals, 'polyps'. Its rich ecosystem hosts many species including starfish, mollusks and octopus. As well as protecting coastal erosion, reefs also act as 'sinks' for carbon, helping slow global warming by absorbing excess atmospheric carbon.

▼ **Ocean role:** Oceans are the most diverse of the marine ecosystems, playing a part in rainfall, oxygen generation and carbon capture as well as sustaining many plants and animals.

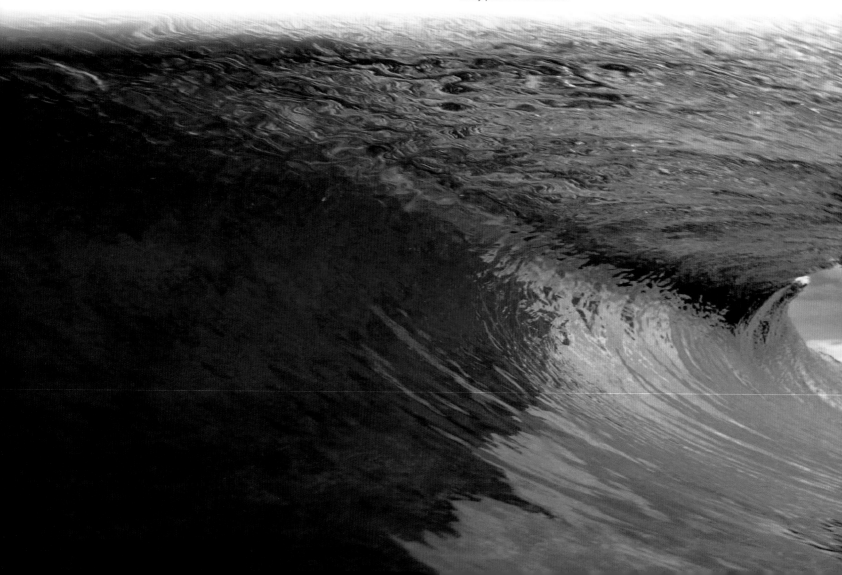

INTRODUCTION

'This we know. The earth does not belong to us; we belong to the earth... Whatever befalls the earth befalls the children of the earth. We did not weave the web of life; we are merely a strand in it. Whatever we do to the web, we do to ourselves.'

CHIEF SEATTLE (1786-1866) OF THE DWAMISH PEOPLE, NORTH AMERICA (ATTRIB.)

Go down to the shore on a calm day and look out across the sea. As the waves roll in gently and draw back, there is a sense of touching eternity: ever changing but ever the same.

Or stand in a mighty forest. On a still day, leaves shimmer in sunlight and breeze. Birdsong echoes. An animal passes silently. The huge trees are motionless, replacing former generations over time that stretches far beyond human existence.

In each case, sea and forest, the peace is actually roiling with complexity and change. In the warm, blue tropical seas, for example, coral reefs protect the beach on which you stand from washing away. These corals – made by millions of tiny animals which filter seawater for food – have evolved over millions of years, and in turn support a range of life so colorful and diverse that they are called the 'rainforests of the seas'.

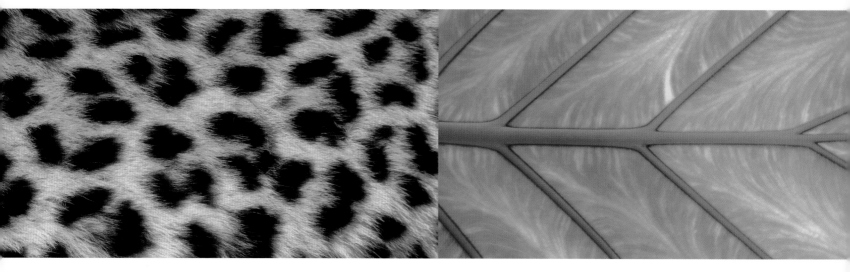

▲ **Changing spots:** People, through hunting, poaching and the spread of agriculture, are the main threat to the big cats. Tourist hunters may pay $30,000 to shoot a lion. The snow leopard, Siberian tiger and Bengal tiger are among the most endangered.

Pages 22-23
Windswept: The Queen's Head rock in Yeh Liu, Taiwan, shows the distinctive wind and wave erosion of the area, still at work today.

Seen it all: Tubu (Teda) people are pastoralists, herding camels and donkeys across the hot, arid lands of northern Chad where the Sahara encompasses the volcanic Tibesti mountain range. The area remains littered with landmines from the 1980s Libya/Chad war; and today is the stage for clashes between the Movement for Democracy and Justice in Chad (MDJT) and Chad's military-backed government.

▲ **Under the microscope:** Green leaves absorb sunlight and convert it to energy in a process called photosynthesis, the base of all plant and animal life – and of the fossil fuels we use.

▶ **Flaring:** Chuhuo is part of Taiwan's Kenting shale, and contains many fissures through which natural gas leaks to the surface and ignites.

The great forest, meanwhile, is a huge green engine, using sunlight to make the sugars that enable plants and trees to grow, and continuously replenishing the oxygen without which all animals would die. Plants and trees are in a constant battle of chemical warfare against attack by viruses, bacteria and animals. But they also co-operate with other organisms in symbiotic relationships such as those with the birds that eat their fruits and spread the seeds far and wide.

These are the 'ecosystems' – self-sustaining collections of organisms and their environment. The biosphere – air, land and water where life exists – has three main ecosystems: freshwater, terrestrial and ocean. Terrestrial ecosystems are themselves divided into seven 'biomes'. These are regions identified by their main plant type, which contains specific plants and animals. The biomes are Tropical Rainforest, Savannah, Desert, Grassland, Deciduous Forest, Coniferous Forest (also called Taiga or Evergreen Forest) and Tundra.

The ecosystems, with their interacting arrays of thousands or even millions of life-forms, have been stable over periods of time that defy human imagination. They are not eternal, of course. For one thing, they are constantly evolving: developing ever more sophisticated and intricate adaptations in the struggle to survive. For another, they have sustained shocks and catastrophic changes over the history of the planet. Deserts replace the richest rainforest. Enormous icecaps have smothered grasslands. Whole seas dry out, and what once was coral reef becomes mountain, three thousand meters above sea level.

But in the last few tens of thousands of years a new force has appeared on the planet – changing it in ways never seen before.

► **Golden arches:** The forces of nature have created this landscape, the greatest density of natural arches in the world. Rock layers reveal millions of years of deposition, erosion and other geologic events. These layers continue to shape life in Arches, Utah, today, as their erosion influences elemental features like soil chemistry and where water flows when it rains.

◄ **'Earth First! We'll mine the other planets later!'** Humans have mined every continent for oil, metals and gems to support ever more wealthy lifestyles, at great cost to the environment. Goldmines in Nevada pumped out more than 2.2 trillion liters of groundwater between 1986 and 2000 – as much water as New York City uses each year.

▼ **Glacial moments:** This vast field of ice on Alaska's coast flows into Glacier Bay. When seen by Sierra Club founder John Muir in 1879 it was a live glacier, a mass of ice advancing into the sea, forming a wall three kilometers long and 80 meters high. It became a key tourist site but in 1889 an earthquake tipped the front of the glacier into the water. Today it has two arms – one 'live', still reaching the sea, and the other 'dead', melting away on land.

Since at least the 'new stone age', or Neolithic period, beginning some 10,000 years ago, humans have shaped the world on which they depend. We learned to burn huge areas to drive game animals and to replenish grassland. We hunted dozens of the largest animal species on most continents to extinction. And with the development of agriculture we escaped the limits of our ecological niche, multiplying in number far beyond any comparable animal of our size.

But the process of wholesale human change, which some scientists now call the 'anthropocene' (the geological era shaped by humans), only really took off with industrialization. Starting with Thomas Newcomen's atmospheric engine in 1712 designed for pumping water out of coalmines, we learned to harness fossil fuels (first coal, later oil and gas) to multiply our will by thousands and millions of times.

No part of the world, from the remotest Arctic tundra to the ocean floor, now escapes human impact. Thousands of chemicals that never existed in nature are ubiquitous. One group, known as polychlorinated biphenyls, or PCBs, can be found in the bodies of polar bears, whales and dolphins at concentrations that endanger their future. Once-fertile land has been turned to salt waste. Plastics are now present in virtually every part of the environment, from soil to seabed, and embodied in creatures of all kinds.

A particularly dramatic example of human impact is the 'hole' in the ozone layer – the stratospheric 'shield' that blocks much of the ultraviolet light from the sun that damages

life on earth. The hole was first noted in the 1980s over Antarctica. The link to a group of chemicals known as hydrochlorofluorocarbons (HCFCs) was quickly established and measures taken that, if sustained and improved, will bring the phenomenon under control. Here was proof that human interference could cause extremely rapid and potentially disastrous change. But here too is proof that wise action can stem or even reverse the most damaging consequences of our own thoughtlessness.

The greatest impact of humanity in the coming century, however, is likely to be climate change resulting from increasing atmospheric concentrations of 'greenhouse gases'. The consequences are complex and in some cases hard to predict. Take three trends among many in three major parts of the world's web of life – oceans, forest and rivers.

Oceans. As the amount of carbon dioxide (CO_2) in the atmosphere increases, so too does the amount that dissolves into the world's oceans. This gradually alters ocean acidity. At a certain point, zooplankton, the tiny shelled creatures near the bottom of the food chain, will no longer be able to make their shells. Their numbers could go into freefall and a fundamental part of the food chain could disappear. The trend is already well under way and scientists cannot be sure that a pivotal moment of collapse is not already inevitable.

Forests. Rising atmospheric CO_2 is accelerating growth of previously undisturbed, old-growth rainforests in central Amazonia – the world's largest forest by far. This is leading to rapid

Pages 26-27
Industrial output: Russia's Norilsk nickel smelter spews molten slag; Russia is the world's largest nickel producer. Mining leaves an indelible mark on landscapes and, as one of the most dangerous occupations, on people's lives. Instead of increased mining, we could re-use or make more use of minerals already obtained.

Pages 28-29
Green and pleasant land: Pre-industrial farmscape, northern England. The human capacity to adapt, to learn to live within limits and to restore ecosystems that seemed lost, should not be underestimated.

Tilting at windmills: People have used the wind's power as a source of energy for many centuries. The ability to harness the wind to drive grinding stones for wheat probably arose in Persia and the knowledge came to Europe with returning Crusaders. These 16th-century windmills in La Mancha, Spain, were immortalized in Cervantes' novel, *Don Quixote*.

▲ 'Tyger! Tyger! burning bright': Could we let the bright tiger light be extinguished? In 1900 there were around 100,000 of these beautiful animals; today there are only 2,000 or so in the wild.

changes in dynamics and species composition. It doesn't necessarily mean an end to the ecosystem as a whole – although thousands of species may die out even as others take their place – but it does show how human activity changes nature even in places furthest away from human settlement. And the trend could give way to something more dramatic within a few decades, as further climate change leads to drought and rapid decrease in forest cover.

Rivers. A warmer climate has already increased glacier melting in the Himalayas by 30 per cent in the last few decades. On present trends, the great rivers of Pakistan, Bangladesh and northern India will run strongly for the next 40 years with the extra meltwater and then die away as the glaciers run out. When the flow does decline, the impact on a region where the agriculture for at least a billion people depends on those rivers will be momentous.

The 21st century may see faster and more dramatic change to the world's ecosystems in the course of just a few decades than has occurred for tens or in some cases hundreds of thousands of years. In a world that also supports six billion human beings, going on ten billion by mid-century, the changes are likely to make the world a much more dangerous place. The biggest threats to ecosystems include war, climate change and global warming leading to melting glaciers, rising sea levels, drought and fires as well as species loss, from tigers to coral reefs.

Flaming Gulf: War's effects on the environment can be catastrophic. An estimated 13,700 tonnes of toxic smoke poured daily into the atmosphere from the hundreds of oil wells set on fire during the first Gulf War. The smoke blew over hundreds of kilometers, inflicting respiratory and carcinogenic damage on those who breathed it in.

Up in smoke: Burning tires in Mexico release highly toxic smoke, polluting the air. How long before every element of the industrial system can be reused, recycled and reprocessed without harm to human health and the natural environment?

▲ **Burn-off:** Traditionally fire has been the main land management tool used by the Aboriginal people of Australia. Here, Darrell, Peter and Frank Milpurrurru start a burn-off at the end of the dry season to reduce the risk of severe bush fire when lightning strikes in November. The burn-off also provides a good opportunity for hunting.

▶ **Omnipotent sand:** Villagers in Mauritania secure sand dunes in an attempt to stave off desertification. The UN Convention to Combat Desertification notes: 'Forced to take as much as they can from the land for food, energy, housing and income, the poor are both the causes and the victims of desertification. In reverse, desertification is both a cause and a consequence of poverty.'

▲ **Out of sight:** Elephants in Kenya. According to the World Conservation Union, about 25 per cent of the world's mammals are in danger of extinction.

▶ **Future (im)perfect:** We have inherited a wonderful world, the product of hundreds of millions of years of evolution and sometimes turbulent change. Now we are in charge and responsible for the future of life on earth. We have inflicted great damage on most of the earth's life systems. Whether we can make a better future is up to us.

The Millennium Ecosystem Assessment, which involved more than 1,360 experts worldwide and published its first report in March 2005, issued 'a stark warning'. 'Human activity' it said, 'is putting such strain on the natural functions of the Earth that the ability of the planet's ecosystems to sustain future generations can no longer be taken for granted.'

But all is not yet lost. The human capacity to adapt, to learn to live within limits, to design sustainable systems, and to restore ecosystems that seemed lost should not be underestimated. Humans have inherited a wonderful world, the product of hundreds of millions of years of evolution and sometimes turbulent change. Now we are in charge and responsible for the future of life on earth. We have inflicted tremendous damage on most of the earth's life systems. Whether we can make a better future is up to us.

If what looks like a calm sea turns to raging fury, it will be at least partly of our own making. But however much is lost, there is a chance that we can learn to navigate through the worst of it. With luck, skill and compassion for fellow sailors, we may yet find a safe haven where we can once again embrace the wonder around us.

Caspar Henderson

PART ONE

Several times in its three-and-a-half billion-year history, life on earth has undergone catastrophic change. The Cretaceous and Permian extinctions wiped out some 90 per cent of animal species. In the long run, life on earth is tremendously resilient, evolving countless wonderful new forms to fill all the niches the planet offers. But in the short term we modern humans – who have only been around for 40,000 years or so – are inflicting one of the biggest extinctions and transformations the planet has ever seen. We need to be aware of our inheritance – and of what we are losing.

& TRANSFORMATION

Pages 38-39
Bora Bora, French Polynesia.
Pages 40-41
Kilauea volcano, Hawaii.

▶ **Lighting up:** Lava is liquid rock, nothing else – but it can burn, as in this image where hot lava sets fire to vegetation.

◀ **When dinosaurs ruled:** Some 120 million years ago, dinosaurs lived here in what is now Paraiba, Brazil, on the banks of a shallow lake, and their tracks have been preserved in the mud. Theirs is probably the most famous extinction, 65 million years ago. Whatever caused their demise – an asteroid collision or a volcanic eruption are among the possible explanations – also caused the death of around 70 per cent of all species on earth. But dinosaur death created a new opportunity – the space in which mammals could develop.

▼ **Life support:** Mono Lake, California, and its watershed contain 14 different ecological zones, over 1,000 plant species and roughly 400 recorded vertebrate species, including sagebrush, Jeffrey pines, volcanoes, tufa towers, gulls, grebes, brine shrimp, alkali flies, freshwater streams and alkaline waters. Around 15,000 years ago most valleys in this region were filled with water. Since the onset of a drier, warmer climate these lakes have shrunk or dried up. They have no outlets, so their waters have varying concentrations of salt and other minerals, creating a rich ecosystem. Millions of migratory birds such as Eared Grebes and Wilson's Phalaropes depend on the lakes for food and refuge.

Life was probably born of fire – maybe in strange forms like today's archaebacteria, 'extremophile' microbes found in conditions that would kill other creatures. More than 80 per cent of the earth's surface is volcanic in origin. The sea floor and some mountains were formed by countless volcanic eruptions, and their gaseous emissions formed the earth's atmosphere. An erupting volcano can trigger tsunamis, flashfloods, earthquakes, mudflows and rockfalls. Nature's capacity to evolve and change is extraordinary. The ancient lakes of Utah and California; the fossilized footprints of a dinosaur; the strangely shaped rocks of Asia Minor (now Turkey): all bear witness to a fascinating, rich heritage.

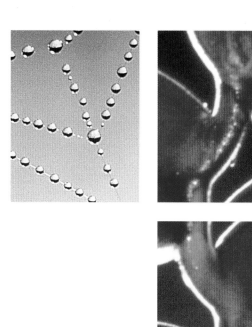

► **Pin-point precision:** Hummingbird in Tanbopata Reserve, Peru.

▼ **Infinite variety:** Over hundreds of millions of years, life has evolved to fill every 'ecological niche' – a specific habitat and pattern of life. The environment is full of such niches, each a potential 'home' for life of which animals and plants can try to take advantage.

Pages 44-45

Stoned: Eruptions of volcanoes three million years ago covered Turkey's Cappadocia plateau with tuff, lava and volcanic ash, leading to one of earth's strangest landscapes. Thousands of years of wind blast and rain erosion on the soft volcanic stone, topped with hardened larva caps, has created a surreal gallery of fascinating shapes and colors.

Dry as dust: Deserts like Namibia's Namib may have been caused by temperature change that killed plants, leading to soil erosion. Wind and water continue to shape deserts, blowing the sand and eroding rocks.

► **All the better to see you with:** Crocodiles – Nile crocodile, right, and in Thailand, below – are a living reminder of an ancient life-form. They have been largely unchanged for hundreds of millions of years. The vertical pupil of their eyes opens wide to give excellent night vision.

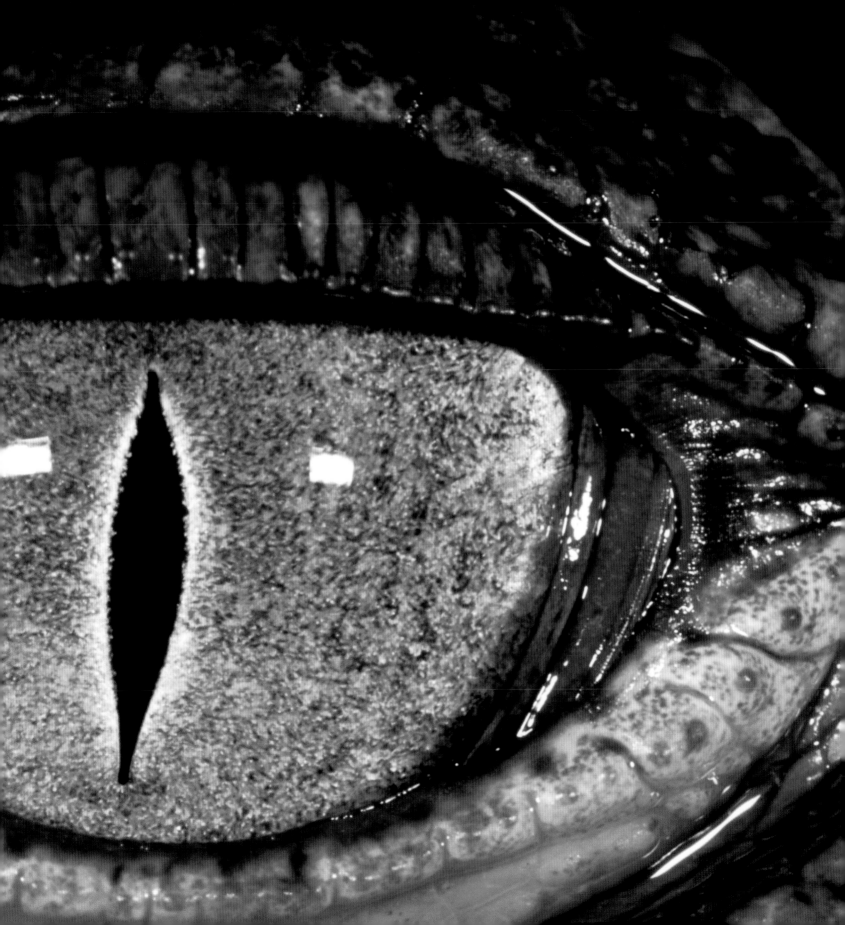

THE SEAS

From time immemorial, humanity has regarded the oceans as a limitless source of bounty. But after two centuries of industrialization and huge growth in human numbers, the seas are in deep trouble. Fish populations are being fished to the point of no recovery. This century may see the near-total destruction of coral reefs – the richest of all ocean ecosystems – by rising temperatures and acidity. Now careful husbandry – including the creation of large off-limits areas – may be the only solution to ensure that our descendants enjoy at least part of the wonderful world handed down to us.

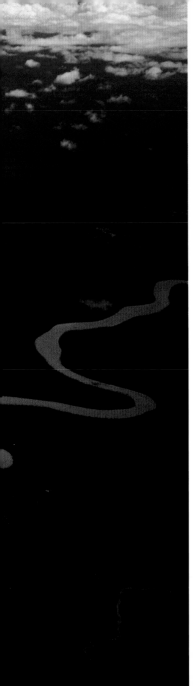

Pages 52-53
Big waves, Hawaii.

◀ **Swamped:** Mangroves occur in around 70 countries in tropical regions from West Papua to the US. They are tidal forests that help prevent erosion and provide a rich habitat for fish and birds. Since 1980, around 35 per cent of the mangroves have been lost to logging and aquaculture.

▶ **Fanfare:** Corals like this giant sea fan are among the first ecosystems that could be completely wiped out by human action.

▼ **Ocean view:** The seas are 71 per cent of the earth's surface. The Pacific alone covers a greater area than all the world's land put together. Phytoplankton, microscopic ocean plants or algae, provide half of the primary production in the world.

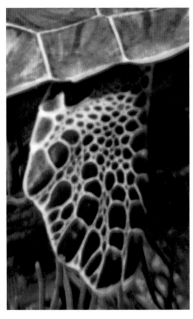

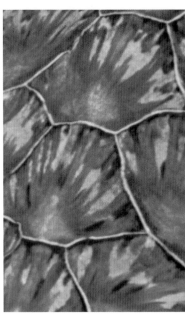

▶ **Turtle treasure:** An ancient Native American belief sees their earth as 'Turtle Island', in which all life is carried on the back of a turtle. Now the world is turned upside down: turtles depend on humanity for their precarious survival. This goes for other precious animals like the tiger, polar bear, whale and orang-utan.

Pages 56-57
Just the spot: Some 130 species of fish like to hang out in and around Australia's Great Barrier Reef at spawning time. Spawning fish need protection from human activity such as fishing and tourism.

Whale tale: Beluga whales, Somerset Island, Canada. Whaling has led to the near-extinction of several species such as the Blue Whale. The ban on commercial whaling has helped some whales recover their numbers. Communities that depend on whaling for food and livelihood are exempt from the ban.

FOREST & SWAMP

Forests and swamps are our ancient home, and a source of endless wonder. They are probably disappearing faster than any other ecosystem on the planet except for coral reefs. Ancient forests fossilized to form the coal that powered the industrial revolution, and made the modern transformation of the planet possible. Swamplands are home to many amphibians such as frogs, which are declining in number partly because they are so susceptible to even low levels of pollution. The remaining examples of 'pristine' woodlands are rapidly eroding, but just holding on – from the pockets of primeval oak forest in Britain to the remnants of deep forest in Southeast Asia that were once home to the orang-utan. We remain dependent upon these environments for fuel, construction, food – and for our dreams.

Pages 62-63
Red-eyed tree frog, Central America.

▶ **Rooted:** The northern coast of South America has important forests and the Orinoco-Amazon mangroves and coastal swamps. The region is an ideal resting place for migratory birds such as scarlet ibises, herons, frigatebirds and greater flamingos.

▼ **Not waving but growing:** Giant ferns like these in Uganda resemble the earliest trees on earth. They originated in the carboniferous era some 300 million years ago, which transformed the climate and laid down the deposits that became coal. These giant ferns probably survived because of their ability to synthesize complex secondary organic poisonous substances and so avoid being eaten by herbivores.

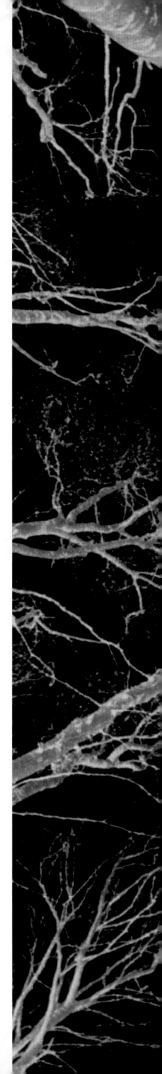

Life force: Plants convert carbon dioxide to organic material and release oxygen. In the 'carbon cycle' – one of the earth's regulating mechanisms – carbon absorbed by plants millions of years ago was transformed into coal and oil.

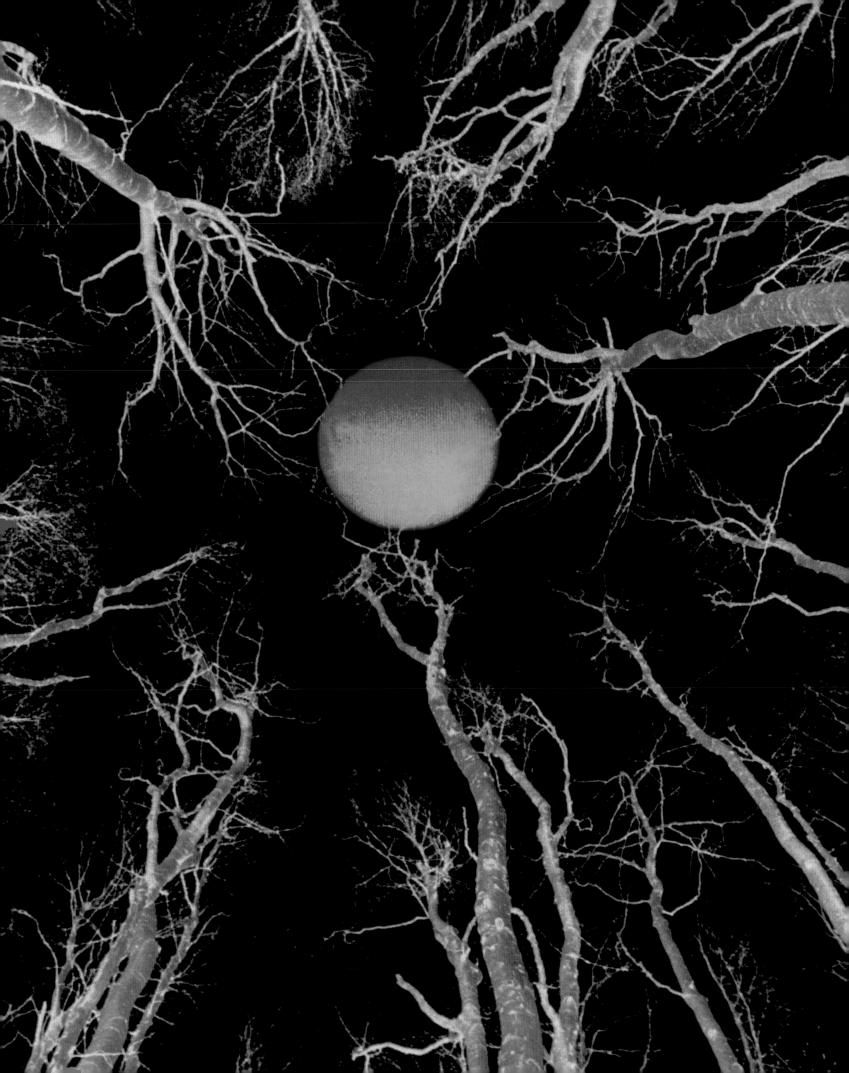

▶ **Wood for trees:** Maples, Japan. The deciduous (shedding leaves) forest biome, between the polar regions and the tropics, has a warm and a cold season with steady rainfall. Much of the world's human population lives in this region. Fallen leaves contribute to soil fertility and this, together with the biome's 5-6 month growing season, makes the region attractive for agriculture – so that many deciduous forests have been cut down.

▶ **SOS:** Logging, forest fires and illegal trade are major threats to orang-utans, here seen in Kalimantan, Indonesia.

▶ **Medicinal bark:** Many trees not only provide food and shelter for animals, but also have medicinal properties for both animals and humans. Knobwood tree, DR Congo.

▼ **Swinging in the rain:** Rainforests such as this one in Sabah, Malaysia, are the foremost ecosystem in terms of species diversity; they cover only about six per cent of the earth's surface yet hold over half the world's species, many still unidentified.

Flight for life: An egret in Japan lifts off. As many as 12 per cent of the world's birds are now threatened with extinction, mainly from loss of habitat. We have, for example, already lost the Atilan and Colombian Grebes, the Wake Island Rail and the Canary Islands Oystercatcher.

MOUNTA

Pages 72-73
Queen Maud Land, Antarctica.

◀ **Northern Lights:** In Finnish they are called *revontulet* – 'fox fires' – which comes from an old story of the arctic fox starting fires or spraying up snow with its brush-like tail. The prosaic version is that the sun gives off high-energy charged particles that travel out into space. When this 'solar wind' of particles collides with the gases in the atmosphere at the magnetic poles they start to glow, producing the spectacle that we know as the auroras – like *aurora borealis*, the Northern Lights, seen here from Norway.

▶ **Hard core:** Extreme environments of ice and rock are most vulnerable to change. Sierra Nevada, California, USA.

They may seem rugged and beyond human reach, but mountain environments and polar regions are undergoing some of the fastest changes as a result of human activity. Every creature in the Arctic and Antarctic bears pollutants such as PCBs from 20th-century industrial processes. The atmosphere above the Antarctic has been transformed by an ozone hole that we have created, now slowly healing thanks to prompt worldwide action. And climate change will utterly transform the landscapes of these most extreme areas: glacial melt, thawing permafrost and disappearing forests mean an uncertain future, with the death of thousands of wonderful species – including perhaps the polar bear – desperate water shortages, and floods.

▶ **Nature the sculptor:** Nature's sculpting of ice, here in Alaska, surpasses any human art.

▼ **Polar bust-up:** By 1970 polar bear numbers had dwindled below 10,000. A 1973 treaty protects their territories and bans their capture – with a few exceptions for native peoples and scientific research. Numbers are now back up to around 30,000. However, global warming is leading to the loss of their habitat as Arctic ice melts, reducing the base of their food chain and diminishing their hunting grounds.

▶ **No ozone:** The discovery of the 'ozone hole' in the 1980s showed beyond doubt that human activity (mainly the use of chlorofluorocarbons in aerosol sprays and fridges) could lead to sudden and potentially disastrous changes in the environment. The ozone layer in the atmosphere protects us from the sun's harmful ultra-violet rays. In 2000, the largest ozone hole ever seen opened up over Antarctica, a sign that ozone-depleting gases churned out years ago are just now coming to their peak. The giant blue blob spreads over 28.3 million square kilometers.

▼ **Earth's pinnacle:** The Himalayas are the world's highest mountain range, formed about 60 million years ago. Even here, life exists. The organisms called extremophiles were only recognized in the 1970s, but can survive in the harshest environments – in the cold of the Arctic and Antarctic; in volcanic vents on land; on the ocean floor; in very dry places; in hot volcanic vents of the deep ocean; in rock, deep inside the Earth; in severe chemical environments harmful to most life-forms; and even in high-radiation environments, such as on the control rods of nuclear-power plants.

Pages 78-79
Mountain high: Mountains often provide rich varieties of habitat at different heights as the vegetation changes. The sandstone range of Alishan, Taiwan, has tropical, sub-tropical and temperate forest – home to the Taiwan macaque, the Mikado pheasant and other endangered species.

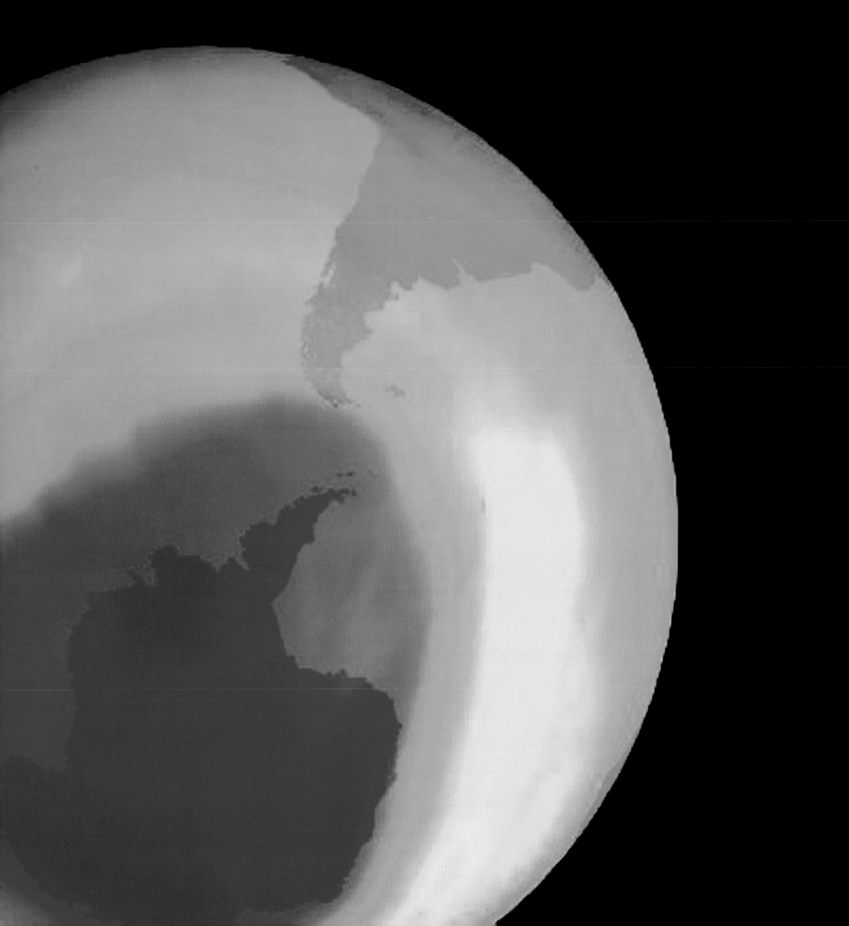

MANAGED LANDS

Pages 82-83
Aboriginals in Arnhemland, Australia.

◀ **Pastures new:** People like these in Mali have moved their herds and flocks to find new grazing and fodder since the domestication of animals began some 10,000-20,000 years ago. This practice – transhumance – has sometimes led to desertification.

▶ **Wild side:** This Xingu boy growing up in the Brazilian rainforest lives in what may seem a wilderness to Western eyes. But what can appear pristine rainforests have often been managed by indigenous people to favor certain sorts of trees with edible fruits or to attract particular game animals.

Humans probably first learned to manage land through fire – burning large areas of savannah to trap game and to improve soil fertility. The practice is still found in parts of the world such as the Australian Bush. Agriculture gave us the ability vastly to increase our population and so to undertake great works of water and soil management. Today the world's most productive land, which once supported a profusion of wildlife, is under human control. Some patterns of land management have increased soil fertility and allowed for the thriving of many species alongside humans. But often land has been catastrophically mismanaged. Salination, excessive deforestation, soil erosion and other abuses have led to famine, war and population crash. Ownership – responsibility towards the land and its future – is one of the fundamental moral and political struggles.

▲ **Nurturing the future:** For thousands of years, people have managed small gardens, cropping intensively, as seen here in Niger. Could this be part of the future as well as the past?

▶ **Steppe by steppe:** Rice terraces are found in the mountainous parts of southern China. The Dragon's Backbone Terraces were built from the Yuan Dynasty to the beginning of the Qing Dynasty. So they have been around for hundreds of years – an early example of human management of the landscape.

▶ **Village people:** Karamojong
pastoralist people in Uganda create
villages from their semi-permanent
settlements with small clusters of
several homesteads. Scarce and
unreliable rainfall means frequent
droughts and a continual search for
water and grazing for their cattle.

Pages 88-89
Greater grass: Yak herd in
Kyrgyzstan. Grassland in its widest
sense includes savannah, prairie,
scrub, high plains and tundra. It
is a recently evolved ecosystem,
covering 30-40 per cent of the
earth's surface. Grasslands provide
grazing for domesticated animals
such as yaks, sheep and cattle.

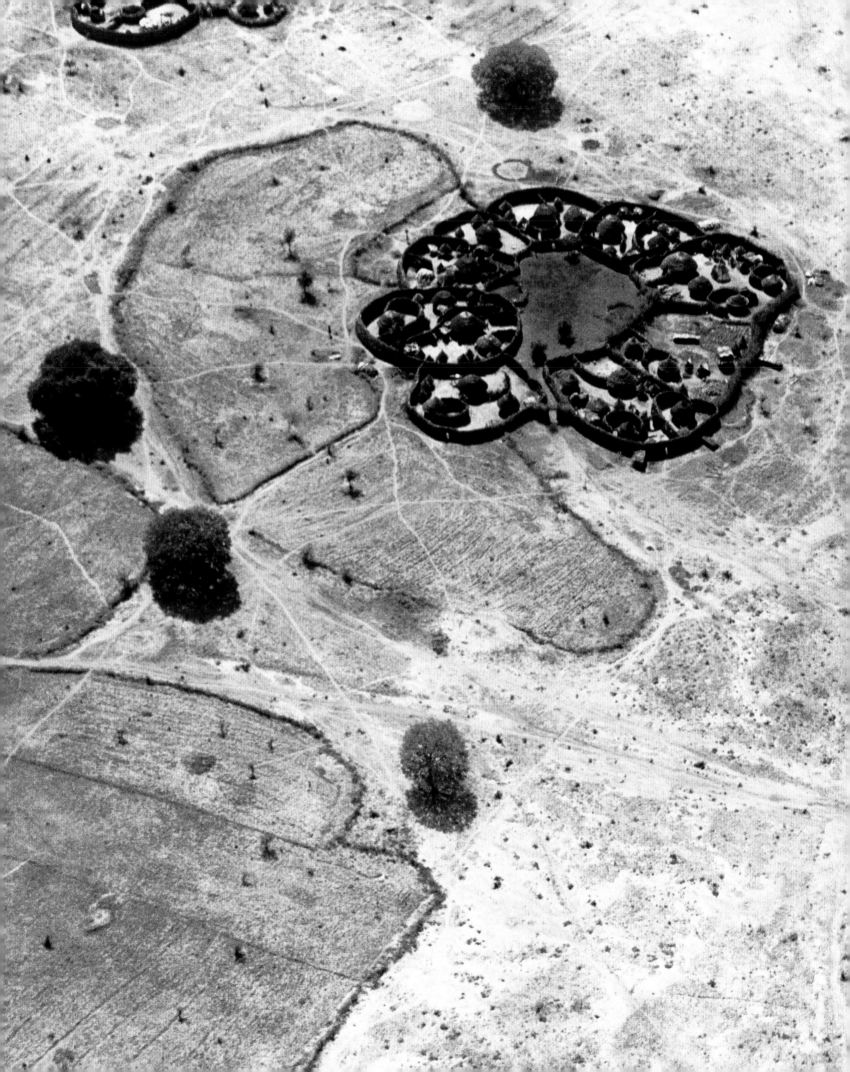

▲ **Land rights:** Masai in Kenya are agitating for the return of land leased to British settlers 100 years ago. The original lease expired in 2004 on one million hectares of land, traditionally used by the Masai but then occupied by white farmers.

▶ **The land is ours:** Members of the landless peasants' Movimento Sem Terra (MST), assemble before occupying the Fazenda Santa Rita. Less than three per cent of the population owns two-thirds of Brazil's arable land. While 60 per cent of the country's farmland lies idle, 25 million peasants struggle to survive by working in temporary agricultural jobs.

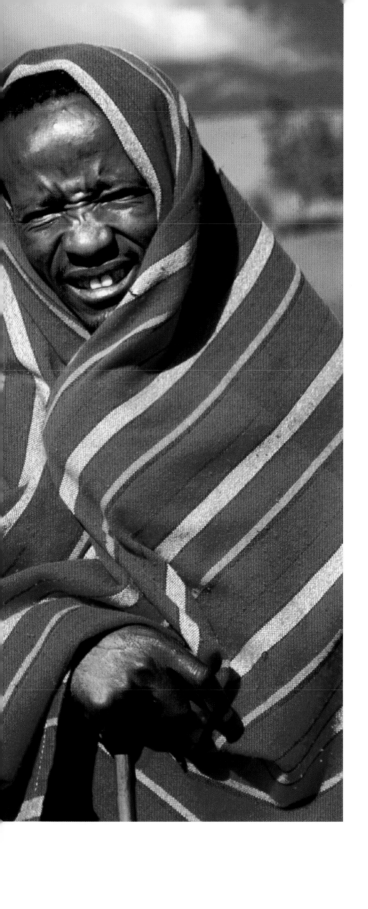

PART TWO
THE LEGACY

PEOPLE

Pages 94-95
Plastic bags, England.
Pages 96-97
Vietnamese children playing.

◀ **The joy of kids:** A migrant family in Brazil. Most women want to have children – but they would like to be in control of when they have them, and how many. An estimated 350 million women still do not have access to contraceptive services.

▶ **Hanging on:** China's population growth rate has stabilized, and at an average 1.8 births per woman is now below the replacement level of 2.1. But it will take decades for this to feed through sufficiently for the population as a whole to remain static or decline. In the mean time there are still 11.5 million more people in China each year.

What kind of world are we making for the future? It will surely be more crowded – up from a little over 6 billion people today to around 10 billion by mid-century. Will this increased human population be able to thrive in a world still rich in natural beauty?

▲ **Way to go:** Integrated programs, looking at family health, education and the environment together, are increasingly the approach to population issues. Yanomami child, Venezuela.

Pages 100-101
Empires of overconsumption: The richest sixth of humanity – which mostly lives in the industrialized countries of the North – has a disproportionate impact on the world's ecosystems through its huge consumption. If each person alive today consumed at the rate of the average US citizen, we would need three more planets to meet the demand. Shopping mall, Kuala Lumpur, Malaysia.

▶ **Taking responsibility:** Father and children in Mozambique. In sub-Saharan Africa, AIDS is the main killer. Some argue its spread is driven by a masculinity that stands in the way of taking responsibility, for example by wearing condoms.

Pages 104-105
Jumping for joy: Vladivostok, Sea of Japan. We can celebrate the 'ordinary miracles' of human endeavor, such as the eradication of smallpox in 1977, which required the co-operation of thousands of people across the world. Other successes are the increase in organic farming, the rise of renewable energy and conservation awareness – but there is still so much to do.

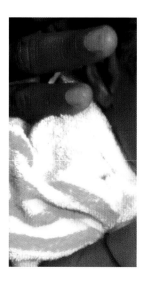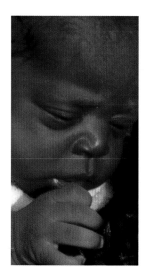

CITY LIFE

Humanity is passing a 'tipping point' in which, for the first time, more than half of us live in cities. Urban living can bring new levels of political awareness, mobilization, freedom and access to resources. But for hundreds of millions in the world's new megacities, it is more likely to mean poor housing, unsafe water and severe air pollution. Urban life – supported as it is by patterns of industrial production that tear at the fabric of life on earth – can alienate us from the natural environment. But making cities sustainable could be the key to the future.

▲ **Recycled dreams:** Garbage pickers in Brazil sifting through the detritus of modern civilization. It is a hazardous occupation, yet it is a common income-generating pursuit for children in many of the world's megacities. And, though there are healthier ways to do it, recycling waste is vital for sustainable cities.

▶ **Wetland no more:** Houses on reclaimed land, California, USA. Draining, filling and conversion to farmlands or cities destroyed an estimated half of the world's wetlands in the 20th century.

Pages 106-107
View of Hong Kong.

▲ **Hot tin roof:** The 17th-century capital of Suriname, Paramaribo, has historic colonial architecture. Today's cities often struggle to create housing that is affordable, let alone attractive; community involvement is vital.

▶ **Growing strength:** Around 800 million city farmers harvest 15 per cent of the world's food supply, according to The Urban Agriculture Network. Food is grown in market gardens and vacant plots, as here in Kathmandu, Nepal, while fish are raised in filtered wastewater.

▶ **Learn from locals:** 'Our professional assumptions of design, construction and managerial superiority were exaggerated. We needed our supposed clients' own knowledge and the skills of local builders.' British architect JFC Turner on helping to rebuild Lima, Peru, after an earthquake.

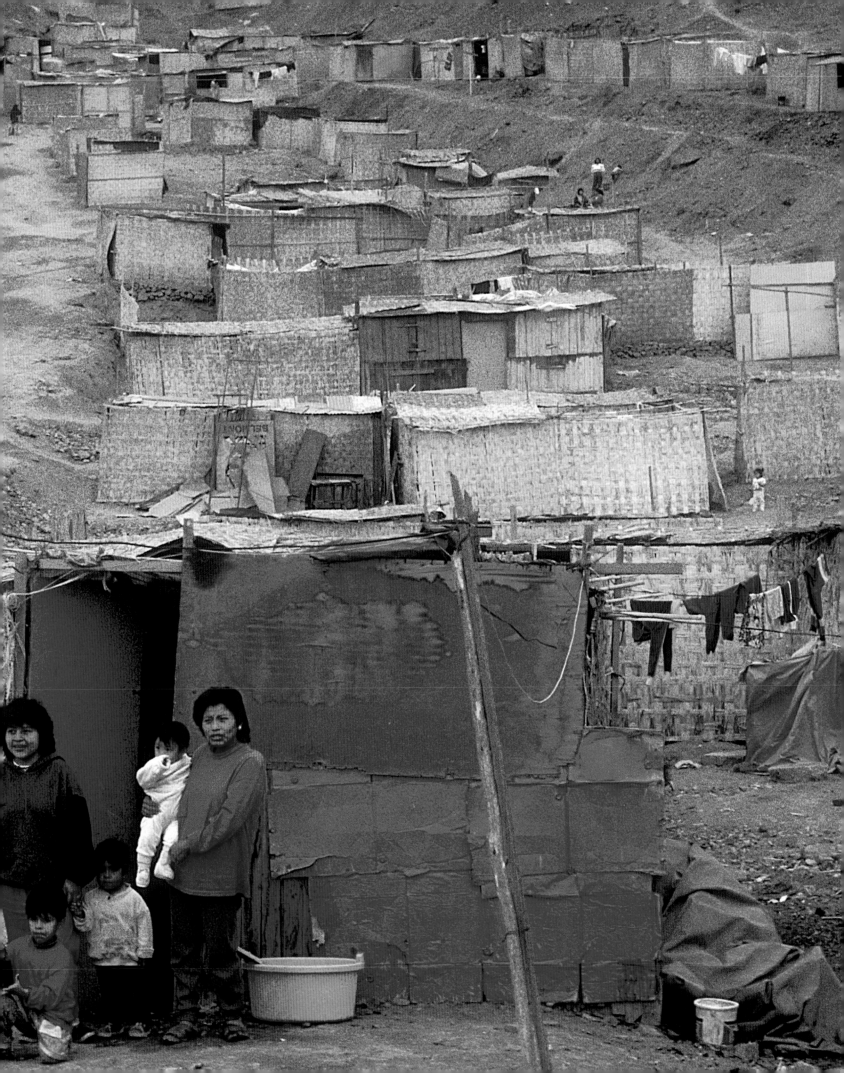

▲ **Urban angst:** Cities such as New York make huge demands on watersheds, fuel and waste processing. Their sprawl can stifle open land and forest. But they have to be part of a future that values both people and nature – for when people live close together, resources can be concentrated.

◀ **Fun in the city:** For some, city life is where it's at. Lots to do, people to see and life on the streets, as here in China.

▶ **Uptown bird:** The red-tailed hawk is the best known and most widely distributed hawk in North America, because of its wider tolerance of habitats – from high mountain country to deserts, from open pastures to urban areas like New York City.

On a carousel: Oktoberfest in Bavaria, Germany. Benefits for both the poorest people and the environment are more likely in cities where there is open democracy and accountability.

116

CLIMATE CHANGE

We cannot know for sure how the global climate will change in the coming decades and centuries. An overwhelming mass of evidence indicates that rapid changes are likely, and that there is a significant chance that the change will be severely disruptive and even catastrophic, especially for the world's poorest people. Vast areas of forest, including the whole Amazon Basin, may die back within a few decades. The rivers of South Asia, on which a billion people depend, may diminish sharply as the glaciers that feed them melt. Even adapting to less severe changes than these will call for tremendous resources and ingenuity.

We have the ability to transform the way we generate and use energy: the next few decades will see some of the greatest breakthroughs since the start of the industrial revolution. But we have to deploy these fast and fairly. If we can, and if we are lucky, we may be able to mitigate some of the more destructive effects of climate change.

Pages 118-119
Firefighter and bushfire, New South Wales, Australia.

▶ **Shelling out:** Around Shell's terminal in Gabon is an important nature reserve, supposedly protected by its national park status. Most people here work for Shell, but the oil reserves are running out. When they do, people may be forced to turn to logging and hunting within the reserve. In 2000, fossil fuels like oil accounted for 77 per cent of world energy consumption. The environmental costs of conventional energy production and use include air, soil and water pollution, as well as acid rain and loss of biodiversity.

▼ **Fuming:** The one-fifth of the world's people who live in the highest-income countries drive 87 per cent of the world's vehicles and release 53 per cent of the world's carbon emissions.

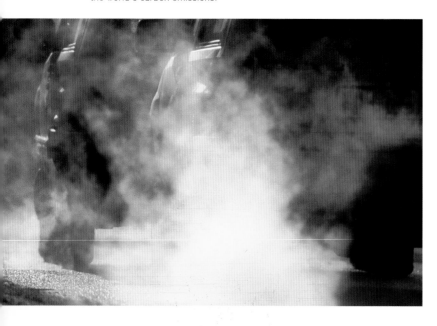

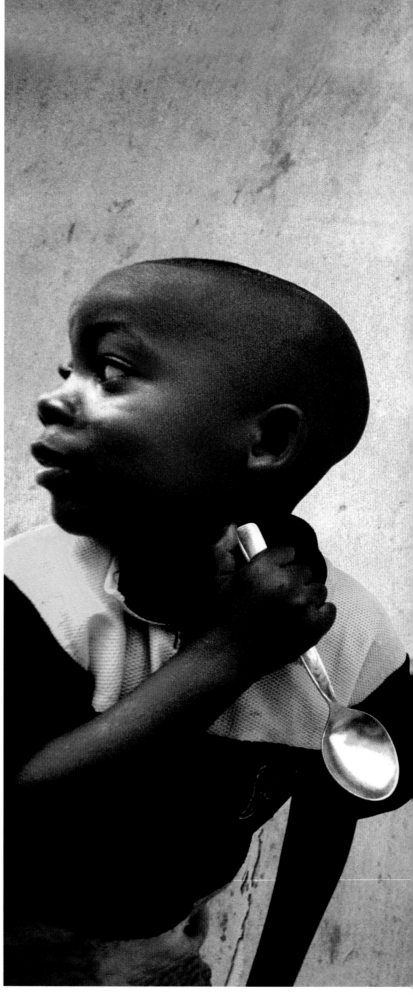

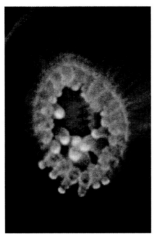

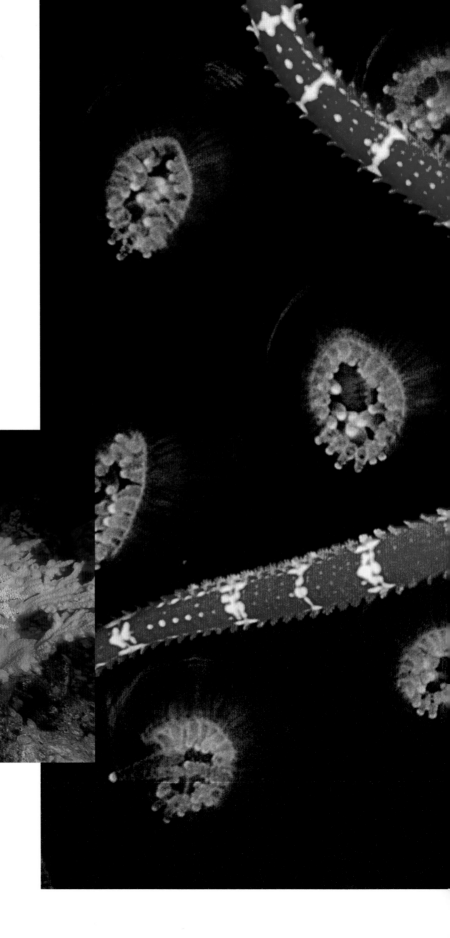

▶ **Perfect harmony:** Ruby Brittle Starfish, in the Bahamas, is one of many species supported by coral reefs. A single reef may hold over 3,000 species of coral, fish and shellfish.

▼ **Bleached out:** Coral reefs, one of the richest ecosystems, are suffering the effects of tourism, overfishing, pollution and disease. The bleaching of coral, seen here in the Maldives, tends to be caused by rising sea temperatures – which themselves are the result of global warming. Some 20 per cent of the planet's coral reefs have already been destroyed with another 20 per cent badly degraded.

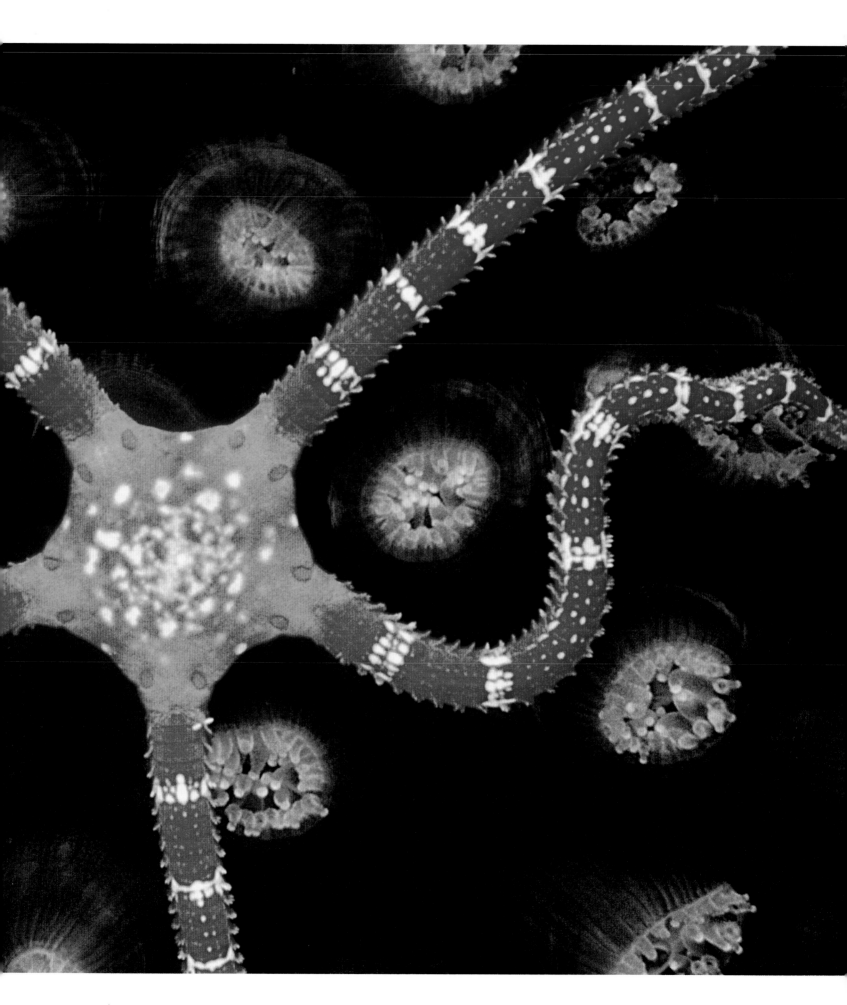

Pages 124-125

Goodbye Tuvalu?: Tuvalu is one of several Pacific island states at risk from rising sea levels, the first effect of which will be to ruin the sources of drinking water. Global warming may cause Arctic and Antarctic ice to melt, thereby raising sea levels and flooding low-lying coastlands.

▼ **Fired up:** Carbon – from burning coal and other fossil fuels – is being released into the atmosphere at record levels: 6.55 billion tons in 2001. Carbon dioxide traps heat, leading to global warming. The US represents 25 per cent of global carbon emissions yet it has even rejected the modest program of emissions reduction envisaged by the Kyoto Protocol.

▼ **Flight from reality:** The Intergovernmental Panel on Climate Change (IPCC) report in 1999 found that aircraft are responsible for 3.5 per cent of greenhouse gas emissions worldwide. Flight is one of the fastest-growing emission sources, and no technological fixes are in sight. Airliner over housing in Hong Kong.

▶ **Ice melt:** Sea levels rise as polar ice melts (here in Greenland), and may submerge low-lying areas like the Maldives and parts of Bangladesh, affecting millions of people, animals and plants.

▼ **After the deluge:** Hurricane Jeanne in 2004 devastated parts of Haiti, killing hundreds of people. Flooding and mudslides have been exacerbated by widespread deforestation as Haiti's people cut wood for charcoal, the island's main fuel.

Pages 128-129
Stormy weather: Arizona, USA. Scientists believe that higher global temperatures will result in more 'extreme weather events' such as storms, floods and droughts. The nine hottest years since records began all occurred in the 1990s and 2000s.

130

▲ **From one extreme...** In South Asia alone, half a billion people irrigate their crops with glacier-fed river flows from the Himalayas. But as the glaciers retreat, the spring meltwater will first surge, causing floods, pictured here in the Bay of Bengal; and then, when the glaciers are gone, stop completely.

▲ **...to the other.** Langtou Gou, a village 130 kilometers from Beijing, is being gradually smothered by sand from the Gobi Desert. Recent sandstorms swept dune-loads of sand into once-fertile lands. Each year, over 100 million people suffer the impact of desertification and another 2,500 square kilometers of land turns to desert. Overcultivation and overgrazing are the main causes. With little vegetation remaining in parts of northern and western China, the strong winds of late winter and early spring can remove millions of tons of topsoil in a single day – soil that can take centuries to replace.

▲ **Taiwan typhoon:** Scooters in the rain during the typhoon season. In 2001, one of Taiwan's deadliest years for storms, Typhoon Toraji killed 200 people. A few months later, Typhoon Nari's huge seas caused Taipei's worst flooding on record and left 100 people dead. In 2004 Typhoon Mindulle killed at least 22 people. Taiwan's National Science Council has openly blamed climate change for the dramatic increase in typhoon activity.

▶ **Fired up:** Each year some 40 per cent of western Arnhemland is hit by bushfires, putting small mammals and plants under threat of extinction. In 2003 Australia experienced its worst bushfires: the National Association of Forest Industries said the greenhouse gas emissions from the fires were equivalent to a year's worth of Australian vehicle emissions. The long-term environmental damage from the fires is often severe, and many lives are lost.

▲ **Nuked out:** Le Port-Boulet power station, France. Nuclear power and fossil fuels have feasted on decades of subsidies; often receiving more subsidies than renewable sources. And as fossil fuels run out, nuclear power may yet make a comeback despite the devastating impact of disasters like those at Chernobyl in the former Soviet Union and Three Mile Island in the US.

▲ **Star fuel in Sudan:** The capacity of photovoltaic cells (PVs) has increased by an average of more than 40 per cent annually since 1992. Japan leads the world in the manufacture and use of solar PVs. Solar cells are already the most affordable option for getting electricity to hundreds of millions of people in developing countries.

▶ **Wind of change:** Wind power is the fastest-growing energy source as concern about climate change coincides with falling costs and favorable government policies. In 2002, it provided enough domestic electricity for 35 million people.

RICULTURE & FARMING

From subsistence vegetable patches in the highlands of New Guinea to the industrialization of big farm agriculture in Europe and the US, humans have transformed more than half the land surface of the planet. Feeding 10 billion people (the likely world population by mid-century) will not be easy. In this century we have to combine the best of the old and the new, increasing levels of nutrition for all, maintaining rural livelihoods and increasing the scope for non-human companions to thrive alongside us.

▶ **Greening Kenya:** Kenya's Green Belt Movement encourages women to plant tree seedlings to avert desertification, and also to grow indigenous drought-resistant food crops such as millet.

Pages 138-139
Wheat farm, Montana, USA.

▶ **Future farming:** The legacy of the 20th century is twofold. On the one hand, there has been a huge improvement in agricultural productivity, meaning that far more people than ever before eat well and live healthy lives. On the other, mechanized agriculture and unwise land-management practices, dependent upon heavily polluting fossil fuels, have introduced vast amounts of enduring toxins into the environment, and have in many places depleted the precious soil which may be the earth's single most valuable and extraordinary resource. Farmland in California, USA.

▼ **Organic works:** Workers harvesting organic rice in India. Organic farming is the fastest-growing section of the world's agricultural economy, from the Philippines to Sweden.

▶ **What's the beef?** Boy and cow in the Philippines. There are currently some 1.3 billion cattle in the world. Large-scale cattle-ranching often leads to environmental damage with the destruction of rainforest and grassland. Since 1960 more than 25 per cent of Central America's forests have been cleared to create grazing for cattle, and over 60 per cent of the world's rangeland has been damaged by overgrazing. Cattle-ranching has also been linked to Global Warming as animals and the industry emit three greenhouse gases – methane, carbon dioxide and nitrous oxide.

▶ **Red hot Dutch:** Noord-Holland province is responsible for more than 50 per cent of the country's total bulb cultivation with a total area under bulbs of 1,168,564 hectares (approximately 30,000 soccer pitches). Today, over nine billion flower bulbs are produced each year, and about seven billion of them are exported, for an export value of three quarters of a billion dollars. The US is the biggest importer.

▼ **Red hot chilli peppers:** Chillies originally came to India from the Americas but today they are widely grown in the subcontinent. Andhra Pradesh is the main chilli-producing state, with 46 per cent of the country's total.

◀ Pages 146-147
Acid yellow: Startlingly bright yellow canola or oilseed rape is grown widely for vegetable oil used in human food and also as animal feed. Particularly in North America, canola is one of the crops most targeted for genetic modification. The hazards for wildlife and humans of such modification are still largely unexplored.

▶ **Boy and beast:** The mainly nomadic Oromo people live in southeastern Ethiopia and northern Kenya, seeking pasture for their herds of cattle, donkeys and camels.

RIVERS & ESTUARIES

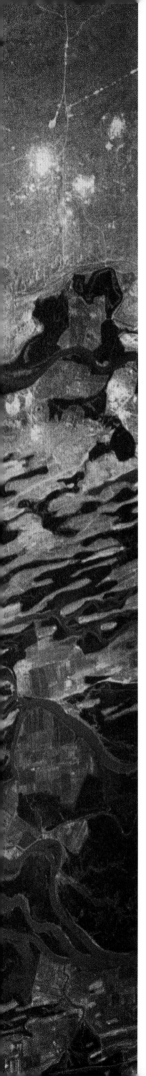

▲ **Pollution solution:** In the mid-20th century, vast stretches of Europe's River Rhine were dead, poisoned by industrial chemicals and wastes. The action taken to clean the river represents a promising feat of environmental protection, and a model for other protection endeavors. Today, with pollution levels down by 90 per cent, salmon and other fish are back.

◄ **Volga clean-up:** At about 3,700 kilometers, Russia's Volga is the longest river in Europe, and one of the most polluted. Some 3,000 factories dump 10 billion cubic meters of contaminated waste into the river every year. Finally some measures have been taken aimed at protecting and restoring the river, including regulation of logging, restrictions on the disposal of industrial waste, and artificial breeding of endangered fish species.

Pages 152-153
Sunset on river, China.

Water – sacred or worthy of the deepest respect in all cultures – is essential to life. Almost everyone lives close to a river, and more than two-thirds of the world's population lives within a few kilometers of the sea. Many rivers and estuaries bear a heavy burden, harnessed almost entirely to human needs – especially for irrigation, which sucks many of them nearly dry, and also as sewers. As a result, river and estuarine environments are among the most polluted areas, degraded and devoid of life. The oceans cover more than two-thirds of the planet and are home to much of the richest and most extraordinary life. The deep transformations wrought by humans – especially in the last century, when much aquatic life was destroyed through overfishing and pollution – have been largely invisible. We are only now beginning to learn the lessons, the key ones being to treat the waters less like a limitless free resource and garbage dump, and more like our own home.

Pages 156-157

Stranded: A beached ship in what remains of Aralsk harbor in the dry bed of the Aral Sea. In 1964 the rivers supplying the sea were pumped dry to irrigate cotton crops – 'white gold' – in Central Asia. Now the water has receded by over 100 kilometers, leaving a legacy of pollution that will severely damage the health of many generations to come.

▶ **Big blue:** Fishing from a bamboo raft on the placid waters of the Li River, amid the classically beautiful Guilin region of southeastern China.

▼ **Something fishy:** Fish farming on a small scale is an old practice providing valuable food. But today's commercial fish farming uses a multiplicity of pesticides and antibiotics. The waste from a large salmon farm can be equivalent to a town of 50,000 people dumping sewage directly into the sea. To feed the fish in these pens requires vast industrial fishing itself, thereby depleting other species. Fish farm in China.

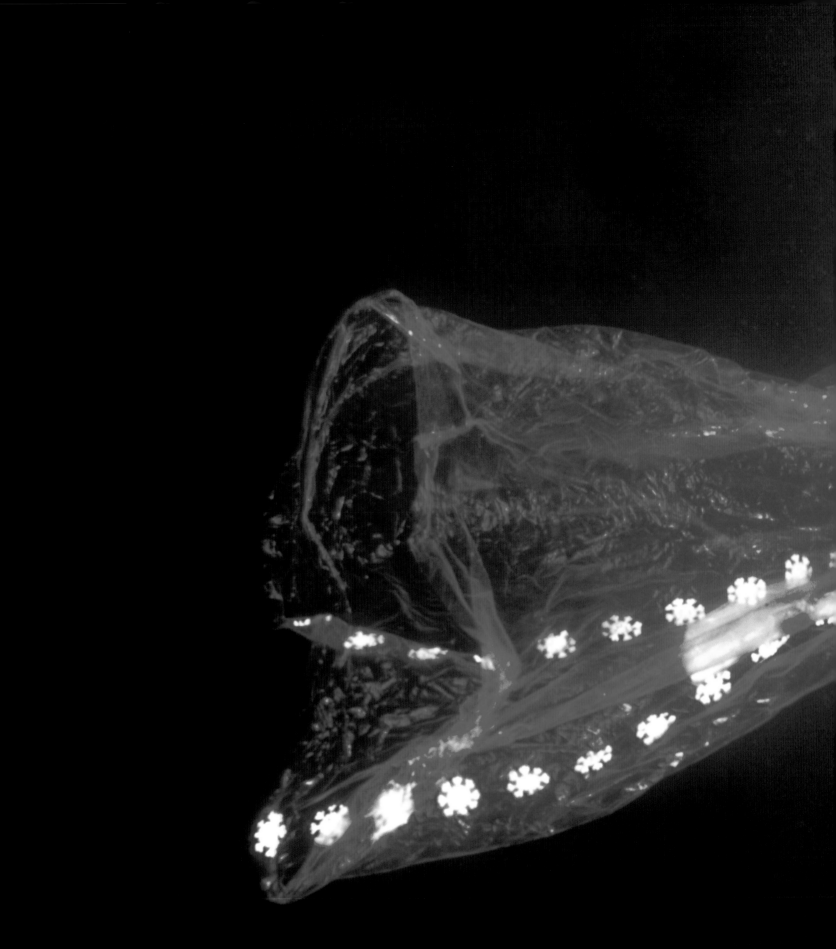

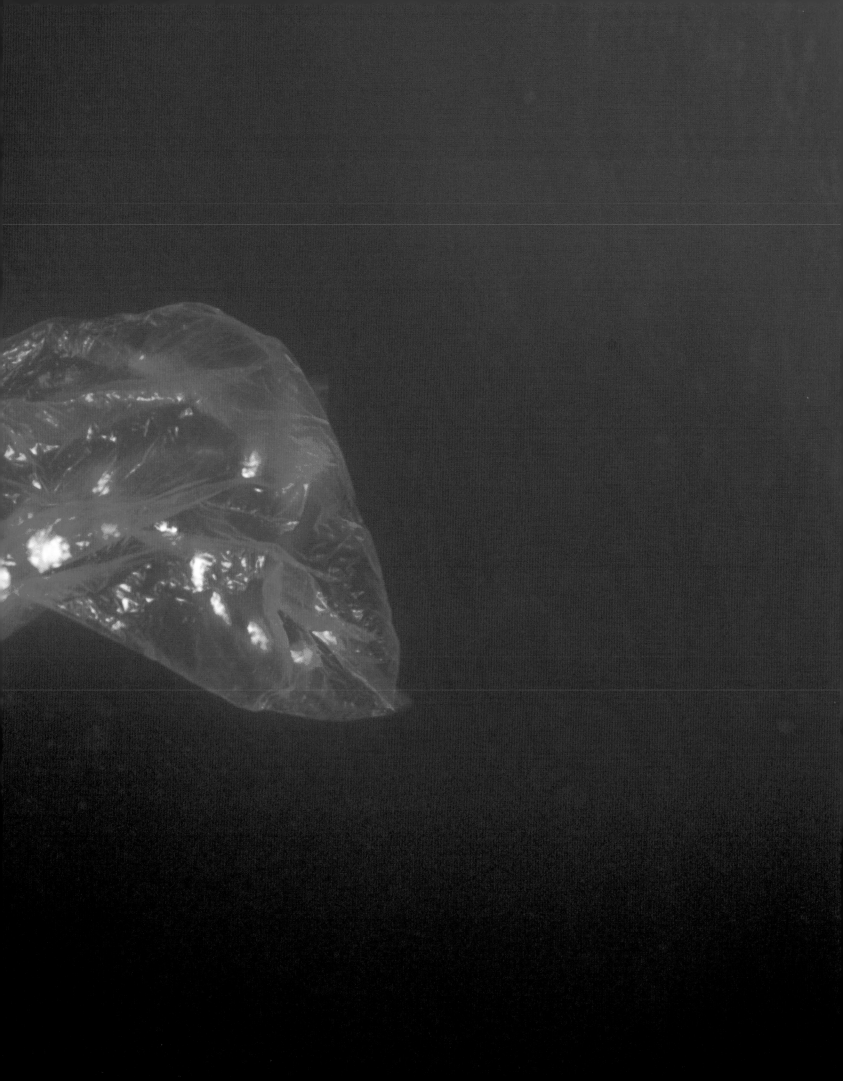

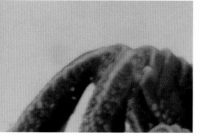

▶ **Seen a ghost?** Or perhaps this ghostcrab in the Seychelles is alarmed over the acidifying oceans.

Pages 160–161
Plastic polluter: Plastic bags, floating here in the sea, and other detritus are polluting rivers and seas. An estimated 50,000 or more synthetic chemicals are in production as plastics, pesticides and other products.

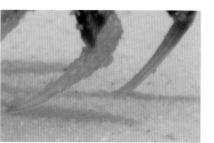

BACK TO THE FUTURE

▲ **Blowing in the wind:** Turbines at San Gorgonia Pass, California, USA. Renewable energy has come of age.
Wind power could lead the way in many countries, generating more than 20 per cent of electricity needs.

▶ **Penguin patrol:** A biologist monitoring penguins in Antarctica – the world's last
great wilderness, and the only continent entirely protected from exploitation.

Pages 164-165
Turtle plans escape.
Pages 168-169
Power of protest: Buddhist monks in Thailand encircle a threatened forest in protest against its destruction.
Wherever you are, organizing against environmental damage can make a difference.

Can humans learn to be true guardians of the planet? Our individuality and relations with each other as people – immeasurably precious as those are – can only make sense if we cultivate our garden earth with more care, leaving space for both wild and wonder. We should be ready for many shocks and surprises. The negative trends – massive extinctions of plant, animal and other species, and the poisoning of the land – will not be reversed altogether in our lifetimes and maybe not even in the lifetimes of several generations hence. In the long run it is not the planet itself that is fragile: natural life may recover many of its former glories long after humanity is gone. But human life on this planet as we know it, now, is as fragile and important as the life of every small child.

▶ **Protecting wildlife:** Young geese from Sweden being guided to the Lower Rhine, Germany, where they will spend the winter. Protecting animals and safeguarding habitats are two key factors for a sustainable world.

▲ **Rivers of change:** By the late 1990s, the river water in Eastern Antioquia, Colombia, was like a cesspool. Cement factories turned it yellow; textile and paper plants made it run blue: it was frothing with chemicals. Environmental authorities tried to clamp down but companies looked for loopholes and greased palms to avoid penalties and stay in business. So instead they used both carrot and stick, applying steadily rising fines to polluters but reducing charges for those that cut their emissions. Some firms began to clean up their act, and one of the largest textile companies announced plans to build a $3-million treatment plant that would recycle the two tonnes of dyes and glues that it had been dumping daily into the region's principal river. The municipal authorities, which were also hit by the charges, put up six new water-treatment facilities. By 2000 organic waste had dropped by 36 per cent. It is at least a start.

Pages 172-173
Gentle giant? As a voracious predator, the orca earned its 'killer whale' name. However orcas are very social, living in pods of 6-40 with close bonds that last for life. The beautiful animal features on totem poles of North America's north west Indians.
Pages 176-177
Striking a balance: The San, like many indigenous peoples, would like to maintain their 'traditional' culture while using what the outside world can offer. 'The Government talks about development. Let it help us with water, then leave us to our own place.' Mogetse Kaboikanyo, who was evicted by the Botswanan Government from the Central Kalahari Game Reserve.

▶ **Biting back:** There are around 10,000 species of bird, 7,000 species of reptile, 5,000 species of mammal and 4,500 species of amphibians in the world today. Habitat destruction – such as logging – is a major cause of extinction. Captive breeding, controversial as it is, may be the only way to protect some animals in the short term. Tiger in Goiania zoo, Brazil.

◀ **Blooming future:** Organic farming is booming, practised in some 100 countries. Almost 23 million hectares are managed organically: 46 per cent of them in Australia/Oceania, 23 per cent in Europe and 21 per cent in Latin America. The market for organic products is growing, not only in the major markets of Europe and North America, but all over the world.

▲ **We hold the key:** Humans have the ability to adapt and to live in sustainable ways – this gives hope for the future. Woman in Niger.

▶ **Don't worry, be happy?** Polar bears evolved at around the same time as humans, and we could go down together. However with luck, skill and compassion, humans may yet fully embrace the wild and wonder around us, and ensure we all survive.

▶ **Damming the future:** China's Three Gorges is the largest hydroelectric dam in the world, constructed to meet the needs of industry, farming and homes as China 'develops'. The dam will displace almost two million people, and threaten the survival of the Chinese alligator, the finless porpoise - the world's only freshwater-adapted porpoise – as well as the Yangtze River dolphin, which only has a few tens of individuals remaining. Work is also starting higher up the Yangtze at Tiger Leaping Gorge, drawing renewed protest from environmental groups and the UN, whose UNESCO agency has recognized the area as part of its Three Parallel Rivers World Heritage Site, a unique center for biodiversity.

◀ **One small step...** Forget men on the moon and rocket science. Responsibility for our planet can start with the smallest things. Something as basic as an energy-efficient light bulb can make a decisive contribution to a brighter future.

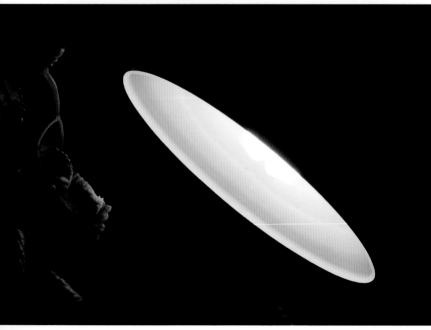

Page 184-185:
Lioness and cub in Masai Mara park, Kenya.

Campaigning organizations

Resources

Photographs and credits

A selection of the many environmental campaigning organizations...

Centre for Science and the Environment – 41, Tughlakabad Institutional Area, New Delhi-110062, India. tel: +91 11 299 55124 email: cse@cseindia.org web: www.cseindia.org

Earth Rights International – 1612 K St NW, Ste 401, Washington, DC 20006, US. tel: +1 202 466 5188 email: infousa@earthrights.org web: www.earthrights.org

Friends of the Earth International – PO Box 19199, 1000 GD Amsterdam, Netherlands. tel: +31 20 622 1369 email: foei@foei.org web: www.foei.org

Greenpeace International – Ottho Heldringstraat 5, 1066 AZ Amsterdam, Netherlands. tel: +31 20 718 2000 email: supporter.services@int.greenpeace.org web: www.greenpeace.org/international__en/

International Rivers Network – 1847 Berkeley Way, Berkeley, CA 94703, US. tel: +1 510 848 1155 email: info@irn.org web: www.irn.org

Oilwatch – Casilla 17-15-24-C, Quito, Ecuador. www.oilwatch.org.ec/english/

Probe International – 225 Brunswick Ave, Toronto ON, M5S 2M6, Canada. tel: +1 416 964 9223 web: www.probeinternational.org

Project Underground – 1611 Telegraph Ave, Ste 702, Oakland, CA 94612, US. tel: +1 510 271 8081 email: project__underground@moles.org web: www.moles.org

World Conservation Union (IUCN) – Rue Mauverney 28, Gland 1196, Switzerland. tel: +41 22 999 0000 email: mail@iucn.org web: www.iucn.org

World Rainforest Movement – Maldonado 1858, 11200 Montevideo, Uruguay. tel: +598 2 413 2989 email: wrm@wrm.org.uy web: www.wrm.org.uy

Worldwatch Institute – 1776 Massachusetts Ave NW, Washington, DC 20036-1904, US. tel: +1 202 452 1999 email: worldwatch@worldwatch.org web: www.worldwatch.org

Resources

Millennium Ecosystem Assessment. www.millenniumassessment.org

The State of the World 2003, Worldwatch Institute/Earthscan, London, 2003.

The Atlas of Endangered Species, Richard Mackay, Earthscan, London, 2002.

Vital Signs 2003-2004, Worldwatch Institute/Earthscan, London, 2003.

'The Death of Environmentalism', Michael Shellenberger & Ted Nordhaus 2004 www.thebreakthrough.org

BBC Education – Range of environmental topics. www.bbc.co.uk/education

City Farmer – Canada's Office of Urban Agriculture. Has links with communities worldwide. www.cityfarmer.org/

Earth Crash Earth Spirit – Explores the environmental, social, spiritual, and health impacts of humans on earth. www.eces.org

Sierra Club – The US's oldest and largest environmental organization. www.sierraclub.org

SolComhouse – Environmental facts and resources on topics such as renewable energy, recycling, rainforests, coral reefs, endangered species, solar power and what you can do. www.solcomhouse.com

UNESCO – The UN's Educational, Scientific and Cultural Organization. www.unesco.org

University of California Museum of Paleontology. www.ucmp.berkeley.edu

Plus information supplied by individual photographers and agencies.

Photographs and credits

vi Kitoga, DR Congo *Sven Torfinn/Panos.*

viii Adelie penguin, Antarctica *Theo Allofs/Corbis.*

x   Mass transit train, Taiwan *Chris Stowers/Panos.*

14 Tassili desert, Algeria *Dieter Telemans/Panos.*

16 Banaue rice terraces, Philippines *Chris Stowers/Panos.*

18 Coral *DiMaggio/Kalish/Corbis*; Wave, French Polynesia *Hank Foto-UNEP/ Still Pictures.*

20 Leopard's fur, South Africa *Roger de la Harpe/Still Pictures*; Alocasia cucullata leaf *M Hazan/Still Pictures*; Natural gas outlets, Taiwan *Chris Stowers/Panos.*

22 Queen's Head rock, Taiwan *Chris Stowers/Panos.*

23 Tubu pastoralist, Chad *Sven Torfinn/Panos.*

24 Miner, Nevada, USA *Bill Varie/Corbis*; Men at Muir Glacier, Alaska, USA *Michael Maslan Historic Photographs/Corbis*; Turret Arch, Arches National Park, Utah, USA *Keith Kent/Still Pictures.*

26 Nickel smelter, Norilsk, Russia *Gerd Ludwig/Visum/Panos.*

28 Windmills at Campo de Criptana, La Mancha, Spain *Günter Ziesler/Still Pictures*; Yewdale, Lake District National Park, Cumbria, England *Martin Bond/Still Pictures.*

30 Tiger on beach *Chase Swift/Corbis.*

32 Tank, Kuwait *John Isaac/Still Pictures.*

33 Burning tire dump, Mexico *Schalharijk-UNEP/Still Pictures.*

34 Aboriginal people, Arnhemland, Australia *Penny Tweedie/Panos.*

35 Sand dunes, Ijnaoune, Mauritania *Clive Shirley/Panos.*

36 Elephants, Kenya *Charlotte Thege/Still Pictures*; Boy in India *Neil Cooper/Panos.*

38 Bora Bora, French Polynesia, Pacific *Truchet/UNEP/Still Pictures.*

40 Kilauea volcano, Hawaii *Douglas Peebles/Corbis.*

42 Dinosaur tracks, Dinosaur Valley, Brazil *DJ Clark/Panos*; Mono Lake, California, USA *Pietro Cenini/Panos.*

43 Lava igniting tree, Hawaii *S&D O'Meara/UNEP/Still Pictures.*

44 Rock formations, Goreme National Park, Turkey *Neil Cooper/Panos.*

46 Dewdrop on web, California, USA *David Cavagnaro/Still Pictures*; Hummingbird, Tambopata Reserve, Peru *Gunter Ziesler/Still Pictures.*

48 Dead trees, Namibia, Africa *Frans Lemmens/Still Pictures.*

50 Crocodile, Thailand *Gavriel Jecan/Corbis*; Nile Crocodile *M Harvey/Still Pictures.*

52 Big waves, Hawaii *Steve Wilkings/Corbis.*

54 Aerial view of mangrove forest, Asmat, West Papua *Bojan Brecelj/Corbis*; Giant sea fan, Fiji *Stuart Westmorland/Corbis*; Aqua Caribbean waters *Neil Rabinowitz/Corbis.*

56 Crescent-tail bigeyes, Great Barrier Reef, Australia *Fred Bavendam/Still Pictures.*

58 Hawksbill turtle *DiMaggio/Kalish/Corbis.*

60 Beluga whales, Somerset Island, Canada *Günter Ziesler/Still Pictures.*

62 Red-eyed tree frog, Central America *Klein/Still Pictures.*

64 Giant ferns, Uganda *Fred Hoogervorst/Panos*; Tree roots, French Guiana *Fred Hoogervorst/Panos.*

65 Oak leaves, Germany *A Riedmiller/Still Pictures.*

66 Sunrise in the forest *BEN/UNEP/Still Pictures.*

68 Forest, Sabah, Malaysia *Fred Hoogervorst/Panos*; Maples in temperate forest, Japan *UNEP/Still Pictures*; Orang-utans in Tanjung Puting National Park, Kalimantan, Indonesia *Cyril Ruoso/Still Pictures*; Thorny bark of Small Knobwood tree, DR Congo *Fred Hoogervorst/Panos.*

70 Egret, Japan *R Kawakami/UNEP/Still Pictures.*

72 Antarctica, Queen Maud Land *Gordon Wiltsie/Still Pictures.*

74 Aurora Borealis (Northern Lights), Norway *Rosing-UNEP/Still Pictures*; Sunrise, Sierra Nevada, California, USA *Galen Rowell/Corbis.*

76 Natural ice sculpture, Alaska *David Muench/Corbis*; Polar bears, Arctic *SJ Krasemann/Still Pictures.*

78 Sunrise on mountain, Taiwan *Chris Stowers/Panos.*

80 Mount Tramserku, Himalayas *Thomas Laird/Still Pictures*; Ozone hole over Antarctica *Reuters/Corbis.*

82 Aboriginals in Arnhemland *Penny Tweedie/Panos.*

84 Family at Hand of Fatima, Mali *Ray Wood/Panos.*

85 Xingu boy, Brazil *Jerry Callow/Panos.*

86 Watering garden, Niger *Jørgen Schytte/Still Pictures*; Rice terraces, Guangxi Province, China *Xintian Pan/UNEP/Still Pictures.*

88 Yak herd, Tien Shan mountains, Kyrgyzstan *Chris Sattlberger/Panos.*

90 Karamojong homesteads, Uganda *Crispin Hughes/Panos.*

92 Masai, Tanzania *Charlotte Thege/Still Pictures*; Movimento Sem Terra (MST), Brazil *Paul Smith/Panos.*

94 Plastic bags, England *Philippe Hays/Still Pictures.*

96 Children, Vietnam *Tran Cao Bao Long/UNEP/Still Pictures.*

98 Migrant family, Brazil *Mark Edwards/Still Pictures*; Mother and child, China *Mark Henley/Panos*.

100 Shopping mall, Kuala Lumpur, Malaysia *Gil Moti/Still Pictures*.

102 Yanomami, Venezuela *John Miles/Panos*; Father and baby, Mozambique *Sean Sprague/Panos*.

104 Jumping into the Sea of Japan, Vladivostok, Russia *Gerd Ludwig/Visum/Panos*.

106 View of Hong Kong *Cornelius Paas/Still Pictures*.

108 Garbage pickers, Brazil *Belerra-UNEP/Still Pictures*; Homes on reclaimed wetland, California, USA *NRSC/Still Pictures*.

110 Wooden houses, Paramaribo, Suriname *Ron Giling/Still Pictures*; Vegetable gardening, Kathmandu, Nepal *Mark Edwards/Still Pictures*.

112 Shanty in Lima, Peru *Ron Giling/Still Pictures*.

114 Street scene, New York City, USA *Jochen Tack/Still Pictures*; Children playing, China *UNEP/Still Pictures*; Red-tailed hawk, New York, USA *Deborah Allen/Still Pictures*.

116 Oktoberfest, Bavaria, Germany *Peter Frischmuth/Still Pictures*.

118 Firefighter amid bushfire, New South Wales, Australia *Dean Sewell/Panos*.

120 Car exhaust *Hartmut Schwarzbach/Still Pictures*; Boy by Shell billboard, Gabon *Sven Torfinn/Panos*.

122 Ruby Brittle Starfish, Bahamas *J Croop/UNEP/Still Pictures*; Bleached coral, Maldives *Pascal Kobeh/Still Pictures*.

124 Fongafale Island, Funafuti Atoll, Tuvalu *Matthieu Paley/Corbis*.

126 Coal-fired power plant *Chris Jones/Corbis*; Airplane over housing, Hong Kong *Chung-Wah/UNEP/Still Pictures*.

128 Approaching storm, Arizona, USA *Gene Rhoden/Still Pictures*.

130 Hurricane Jeanne floods, Gonaives, Haiti *Daniel Aguilar/Reuters/Corbis*; Ice cap breaking into the sea, Greenland *James L Amos/Corbis*.

132 Flooding near the Bay of Bengal *Trygve Bolstad/Panos*; Desertification in Langtou Gou, China *Mark Henley/Panos*.

134 Scooters in rain, typhoon season, Taipei, Taiwan *Chris Stowers/Panos*; Aboriginal children in Arnhemland, Australia, stamp out a bushfire *Penny Tweedie/Panos*.

136 Le Port-Boulet nuclear power station, France *Adam Woolfitt/Corbis*; Solar power, Sudan *Hartmut Schwarzbach/Still Pictures*; Girl and windmill *Olivia Droeshaut/ Rep/Still Pictures*.

138 Wheat farm, Great Plains, Montana, USA *Alex A Maclean/Still Pictures*.

140 Woman with tree seedling, Green Belt Movement, Kenya *William Campbell/Still Pictures*.

142 Organic farming, Andhra Pradesh, India *Mark Edwards/Still Pictures*; Industrial agriculture, California, USA *NRSC/Still Pictures*.

144 Boy and cow, Philippines *Erwin T Lim/UNEP/StillPictures*.

146 Canola/oilseed rape field, UK *Martin Bond/Still Pictures*.

148 Chilies, Lambada, India *Sean Sprague/Panos*; Dutch tulips, Holland *Fred Hoogervorst/Panos*.

150 Oromo boy with donkey, Ethiopia *Martin Harvey/Still Pictures*.

152 Sunset on river, China *UNEP/Still Pictures*.

154 View from space shuttle of Volga delta *NASA/Corbis*; Pike in River Rhine, Germany *Michel Roggo/Still Pictures*.

156 Ship cemetery, Aralsk harbour, Kazakhstan *Paul Howell/UNEP/Still Pictures*.

158 Fish farming, China *UNEP/Still Pictures*; Raft floating on Li River, China *Keren Su/Corbis*.

160 Plastic bag in the sea, Tanzania *Pascal Kobeh/Still Pictures*.

162 Ghostcrab, Seychelles *Jochen Tack/Still Pictures*.

164 Turtle plans escape *Puricello Carlo/UNEP/Still Pictures*.

166 Wind power, California, USA *Martin Bond/Still Pictures*; Penguin, Antarctica *Birnbau/UNEP/Still Pictures*; Biologist on Cape Crozier, Antarctica *Galen Rowell/Corbis*.

168 Buddhist monks protesting, Thailand *Boonsiri-UNEP/Still Pictures*.

170 Orca, Washington, USA *Ray Pfortner/Still Pictures*.

172 Animal protection, Xanten, Germany *Ruediger Fessel/UNEP/Still Pictures*.

174 Children in polluted river, Colombia *German Castro/Panos*; Tiger, Goiania Zoo, Brazil *Weimer de Carvalho Franco/UNEP Still Pictures*.

176 San herder, Botswana *Paul Weinberg/Panos*.

178 Organic wheatfield, France; *François Gilson/Still Pictures*. Woman, Niger *Ron Giling/Lineair/Still Pictures*.

180 Polar bear, Arctic *N Asanis/UNEP/Still Pictures*.

182 Apollo 11 astronaut Buzz Aldrin on Moon, 1969, *photo by fellow astronaut Neil Armstrong/1996 Corbis/Original image courtesy of NASA/Corbis*; Energy-efficient lightbulb *Troth Wells*; Site of Three Gorges Dam, Anhui Province, China *D Stanfill/UNEP/Still Pictures*.

184 Lioness and cub, Masai Mara, Kenya *Nicolas Granier/Still Pictures*.

www.panos.co.uk
www.stillpictures.com
http://pro.corbis.com

About the authors...

Troth Wells joined the NI in 1972, helping to launch the *New Internationalist* magazine and build up its subscriber base. She now works on the editorial team as Publications Editor and has produced five food books including *The World in Your Kitchen* (1993), *The Spices of Life* (1996) and *The World of Street Food* (2005). In addition she is the English-language editor of *The World Guide*, produced by the Third World Institute in Uruguay. She has travelled in Central America, Africa, India and Southeast Asia.
www.newint.org

Caspar Henderson is a writer on environment, energy and political affairs, and an editor at openDemocracy.net, a project dedicated to open global politics. He keeps a blog at http://jebin08.blogspot.com/

About the New Internationalist

The New Internationalist (NI) is a not-for-profit co-operative based in Oxford, UK, with associated offices in Adelaide, Australia; Toronto, Canada; Christchurch, New Zealand/ Aotearoa; and Dublin, Ireland. Founded in 1972 with the backing of Oxfam and Christian Aid, the NI has been fully independent for many years. It publishes the *New Internationalist* magazine, which reports on global issues, focusing on the unjust relationship between rich and poor worlds. **www.newint.org**